Occasional Paper No.29

GW00725694

Buildability – the construction contribution

C. Gray, MPhil, MCIOB

The Chartered Institute of Building,
Englemere, Kings Ride, Ascot, Berkshire, SL5 8PJ

CONTENTS

Buildability – the construction contribution

by C. Gray MPhil, MCIOB

INTRODUCTION

The idea for the research described in this paper grew out of experience from preparing estimates for a wide range of building projects. The particular methods of analysis used were designed to discover the potential construction problems within a particular project, to assess adequately the implications and to cover the risk involved. Over a considerable period and a large number of projects a noticeable pattern began to emerge.

Firstly, in the majority of projects the construction period was either too long or too short compared with a carefully calculated programme, based on normal and appropriate construction methods.

Secondly, details within the designs were very difficult, and in some cases impossible to construct. On interview it was clear that the majority of design teams had not considered the problems which the constructor had to face.

From an estimating viewpoint it could be concluded that had these things been considered, the design altered and the contract programme developed more carefully, then the client would have had a cheaper building.

It was evident, therefore, that the contractor had practical skills which could benefit the industry's clients if only they could be incorporated in the design. The practical problem was how to do it.

Initially, the problem has been considered in regard to the competitive bid type of contract, although it is recognised that there are other types of approach, some of which attempt to incorporate the contractor's experience into the design team. But even in these systems, particularly in the UK, the contractor's role commences too late during the design process to have the major impact that, in theory, it could have. It was evident when talking to design teams that they were disappointed and disillusioned with the response from contractors who were asked to review designs and to identify inbuilt building problems. Their view as to why it was difficult for contractors to contribute was also interesting because it began to reveal the problem which contractor's face.

Whilst the building difficulties are real and apparent, to remove them one has to consider the problem in a conceptual way, a way which is natural for a designer but not for a contractor.

To be successful, the conceptual assessment must be done with a full understanding of the implications of design and of construction. It is the enormous complexity of the interlinking causes and effects at the detailed level which makes the construction problem so difficult.

The problem of evaluating the operational problems inherent in a design is, therefore, complicated and designing a procedure which can make this a matter of routine is extremely difficult.

THE PRINCIPAL CONCLUSIONS OF THE STUDY

'The only way is for the designer to become humbler and go back to the basic understanding of materials and construction processess'

ENZO PIANO 1981

- ● **Construction expertise can make a significant contribution to design**
  There is a growing awareness that the contractor can make a valuable practical contribution to certain stages of the developing design of a building, and that the resulting simplification of the site work will produce cheaper and quicker construction or both. Unfortunately, the actual benefits are extremely difficult to measure and only isolated examples have been recorded. It is possible to extrapolate a potential cost benefit to the UK industry's clients, as a whole, of between a 1% and 14% reduction in capital cost if advice upon the practical aspects of constructing a building are incorporated into the design thinking.

3

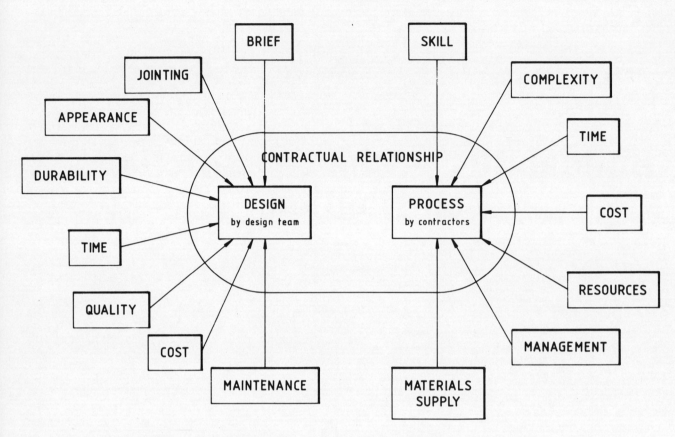

Figure 1 Demands placed upon the design and construction process.

● **The construction contribution must be made very early in the design.**

The maximum contribution of skills should be made from the initiation of the design, within stages A and B of the RIBA Plan of Work, to obtain most benefit. Most contractor selection procedures are such that the design is unable to benefit from an early involvement of the contractor's skills. Therefore, systems need to be developed to include and reflect the input of these skills during all design phases, but particularly during the initial phase.

● **The design process is very complex.**

The hypothesis that there is a methodology which covers the whole problem of analysing the design/construction relationship is probably true but the problem is wide and complex. To provide sensible answers, the implications of the whole design/construction problem must be considered. The designer (Figure 1) has many demands, which are often conflicting, and which need to be balanced and considered when developing the design. The construction process is only one of them. There is, therefore, a need to be able to establish the effect of all the demands placed on the designer when evaluating alternative strategies. Generally, design teams have the information available on the left of the diagram in Figure 1, but not the implications upon the process of construction. This process is also complex and has may interrelated components which must be considered together when alternative strategies are to be considered.

● **Design analysis within the framework of competent management.**

It is possible that a methodology may be developed but the technical problem requires discrimination and well directed analysis. The first stage must be to separate those factors which are external influences on the construction processes as these are invariably man made. Late delivery of components, poor information and late instructions are management problems. The second stage is to examine the work tasks, determine the resource input and for management to ensure that the necessary skills, materials and plant are available. It is only by making the assumption that the project will be built within a framework of competent management that any sensible prediction of the consequences of alternative design strategies can be developed. Once that assumption is made then everything must be organised by the design team which enables the contractor's management to achieve the predicted production. One important implication of this view is that the contractual framework must be considered very carefully to ensure the smooth running of the construction process.

● **Simplification of the sequence of tasks has the greatest return.**

Whilst there is a benefit if the task to be performed by the operative can be simplified by reducing its time consuming aspects, the larger benefits arise if the whole operation can be eliminated. In many cases this is impractical but if operations which interrupt the flow of work by one skill could be removed, then that skill could continue more productively. The significance of simplifying the sequence of tasks is that there is a fourfold return compared with the simplification of the task. Improving productivity in the UK construction industry must first be done by determining whether the proposed design contains a simple or complex sequence of tasks or work packages. By amending the design the sequence can be changed to produce a simpler sequence of work packages, which can then be examined to achieve simplification within themselves. In this way the increase in productivity will be led by the design but in practice this will only take place if the implications can be analysed and evaluated.

- **A methodology for analysing the construction implications of a design.**

Central to all the above requirements is the clear identification of the tasks within the construction process which has led to the development of the methodology of analysis shown in Figure 2. This is based on the concept that a design is a composition of known categories of components and construction practices which can be joined together in a limited number of ways to achieve an infinite variety of buildings for a very wide range of uses. The key in analysing any design is to establish which particular components are to be used, their particular arrangements and their combination. However, it is not necessary to go into the extreme detail of the operative's task, an intermediate level termed the work package is sufficient. Each significant work package can be identified using simple selection criteria.

The complexity of the combinations of the tasks can be determined by identifying the relationship and interrelationship between work packages. The sequences and relationships of these work packages can then be built into models of the design, upon which simulations of the construction process can be made. A method of modelling the complete network of interrelationships has been proposed from which information can be extracted as input to the cost calculations of the consequences. Each part of this method is complex and it must be appreciated that to attempt to model the total implications without a simplified approach would be inordinately expensive and time consuming. But by taking realistic judgements about the important aspects of a project, reasonable predictions can be made.

- **The use of simulation models.**

There is no way of predicting precisely and absolutely the course of events, particularly in anything as complex as construction, even on the day of the actual event. It must be accepted, therefore, that any prediction must attempt to be as accurate as possible. Since each construction project is unique it is not possible to use conventional generalised simulation techniques because the nature of the model cannot be identified until design is underway. Models must, therefore, be built on the solid foundation of

the understanding of good practice to predict the variations that will occur in practice.

- **Conclusion**

The research has revealed that there is not a simple answer to the problem of evaluating the construction implications of a design. The construction process is extremely complex and there is no one best way in which it can be analysed. To be a considered analysis of design, the consequences of the design must be understood from several points of view. It can either be by: the time to construct, the cost of construction, the sequence of operations or a combination of them all. However, the analysis must be acceptably accurate and work within the time frame of the design team's thinking in order to be influential.

To say how this will be done is perhaps speculation but a methodology which is simple, but composed of complex parts, has been proposed, drawing upon recent studies of the problems of the industry. It is the common fault of research that it leads to more research and this subject is no different. Indeed, in this case there are two parallel paths to tread.

First, the research to define and understand the operation and limitations of the basic tasks of the industry.

Second, the development of the modelling systems from the techniques already researched, into useful tools which are simple to use.

The final question which has to be asked is who will use the techniques when they are developed into operational tools?

The design team
A tool will be available to analyse the construction implications of a design and work within the strictures of the conventional contractor procurement procedures. Whether it will be used by the designer directly through a computer aided design system, or by a quantity surveyor as part of the cost planning procedure, depends on the pace of computer software development.

The contractor
When operating within the alternative contractual arrangements such as a package deal, management contracting or project management, the contractor will be able to improve his professional understanding and thus his contribution by a greater conceptual analysis of the project, during the vital design phase.

There is no substitute for construction experience but only when coupled with an analytical capability and perception can it in practice influence the design. Within current UK contractual procedures there is a great need for designers and contractors to understand the whole process and to use each

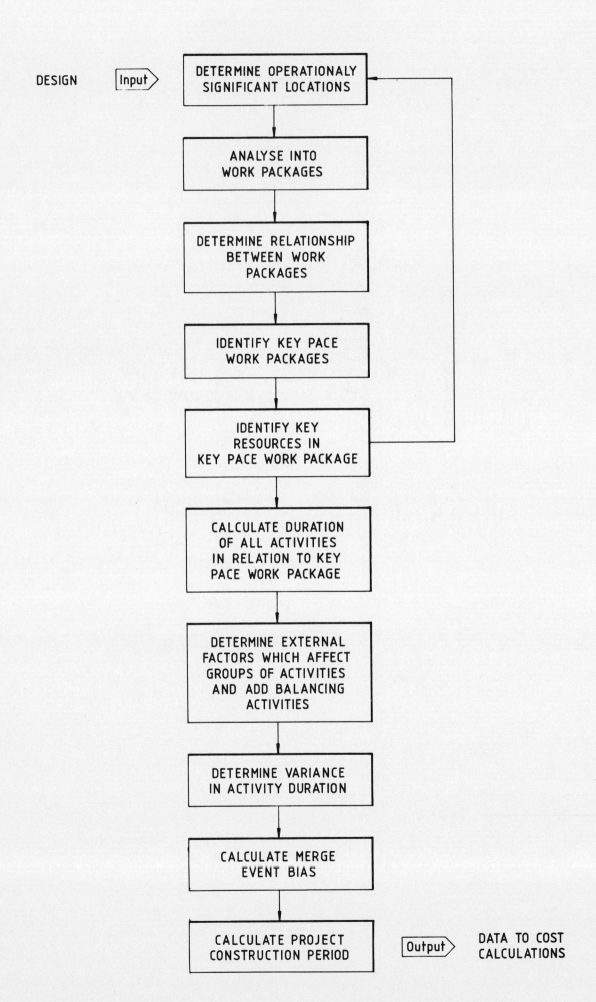

DESIGN   Input⟩

DETERMINE OPERATIONALY
SIGNIFICANT LOCATIONS

ANALYSE INTO
WORK PACKAGES

DETERMINE RELATIONSHIP
BETWEEN WORK
PACKAGES

IDENTIFY KEY PACE
WORK PACKAGES

IDENTIFY KEY
RESOURCES IN
KEY PACE WORK PACKAGE

CALCULATE DURATION
OF ALL ACTIVITIES
IN RELATION TO KEY
PACE WORK PACKAGE

DETERMINE EXTERNAL
FACTORS WHICH AFFECT
GROUPS OF ACTIVITIES
AND ADD BALANCING
ACTIVITIES

DETERMINE VARIANCE
IN ACTIVITY DURATION

CALCULATE MERGE
EVENT BIAS

CALCULATE PROJECT
CONSTRUCTION PERIOD

Output⟩   DATA TO COST
CALCULATIONS

Figure 2 Methodology for determining an effective construction process simulation model

other's strengths, but unless the practices become very direct in the American and Continental way, systems must be developed to aid each in the understanding of the practices of the other.

DEVELOPMENT OF THE CASE FOR DESIGN/CONSTRUCTION EVALUATION

Since the Second World War there have been many studies of the British construction industry, mainly as a result of the disappointing performance of the industry when compared with that of other countries. Each has stated in its conclusions that major change is required. Whilst there is no denying that the construction industry has not remained fixed it cannot change overnight to an entirely new structure and way of working. Change has to be evolutionary rather than revolutionary and must, in the short and medium term, take account of the skills which exist within the industry.

The message which emerges from all the analyses and reports, is that there is a very sharp division between the design team and the construction team. This has persisted in the face of heavy criticism in every study. The UK contract system which is embodied in the JCT form of contract does not state that this will be so, but it is based on the clear implication of a clear division. In 1966 the Tavistock Institute[1] reported a major study of the industry's communication system which recognised the organisation pattern of this division (Figure 3).

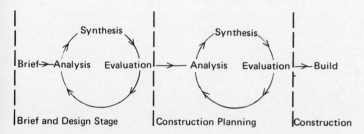

Figure 3 Generalised information flow in the open tender form of the building process.

There was concern that this organisation pattern allowed communication to work in one direction only;

Thus while design affects construction planning, construction planning cannot affect design'.

Looking at nearly every other manufacturing industry, the product designer always works within the manufacturing organisation, because the design has to be manufactured on the production equipment available or obtainable. There is, therefore, a two-way development process between design and production engineering. A manufacturer is also able to produce trial samples, make pre-production runs and tests to enable the design to be refined. The construction industry, however, is an industry which manufacturers to order, to individual designs. Tavistock not surprisingly inquired;

'Whether the information flow of the construction planning sequence might not have had

relevance to design; might in fact have modified the choices made in design'.

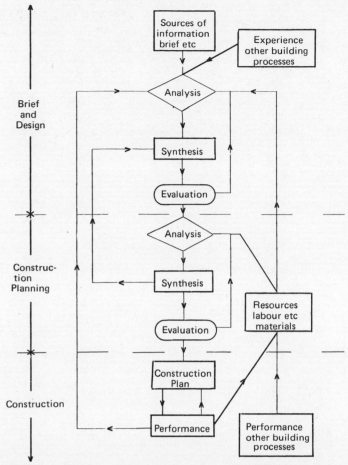

Figure 4 Information flow in an idealised system.

In manufacturing this question is so obvious that it is never asked.

Therefore, the question which might be asked is; if this is beneficial for the construction industry, why has it not changed to take advantage of the benefits? There are two aspects which must be considered in answer to this question. Firstly, the organisation structure required to allow this to happen shhuld ideally be that in Figure 4, which shows a feedback loop system, requiring an evaluation of the design made by the construction team and subsequent revision and development made by the design team. Given the structure of UK contracting, the structure in Figure 5 may be the only one which can be developed, but it does not give the same immediacy of response, because of the diversity of skill and relationships involved.

Secondly, there has been no demonstration of the benefits that this closer linking of design and construction might give and there are many who consider that such benefits are illusory.

Barrie and Paulson[2] in the USA were concerned with the pattern from another viewpoint, that of the inter-relationships between engineering design, construction and operating costs for a facility and the way the level of influence of the costs of a facility decrease as the project evolves.

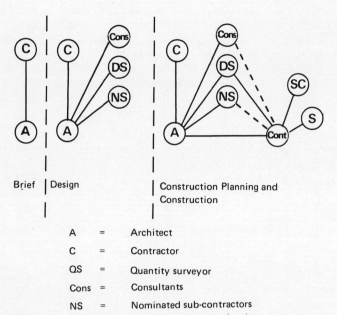

Brief | Design | Construction Planning and Construction

A   =   Architect
C   =   Contractor
QS   =   Quantity surveyor
Cons =   Consultants
NS   =   Nominated sub-contractors

Figure 5 Evolving organisation pattern in the divided case.

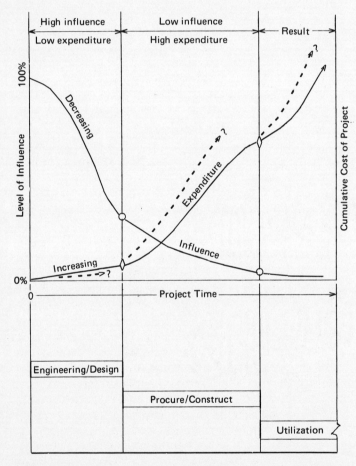

(Source Professional Construction Management, Barrie and Paulson. C. 1980 McGraw Hill.)

Figure 6 Level of influence on project costs.

Figure 6 shows that a high proportion (c.70%) of the total project costs, are determined at the completion of the early design phase, because as the design is developed decisions are made which lock the design into a certain set of relationships. These relationships become so complex that any subsequent change is virtually impossible or cost insignificant. Therefore, if there is any benefit to be derived from the construction

planning contribution that contribution must be made from day one of the design process. The nature and extent of the contribution, however, has not been tested to any significant degree in the UK because there is no method, within the traditional contract system, of the contractor interfacing so early in the design.

To examine whether there is a case for information on the construction process making a significant contribution to the design, a detailed search was made of the published literature. The search has been conducted in three areas, because of the different organisational features of the construction industries in particular parts of the world.

Practice in the United States
In the United States, the usual approach adopted is that of the extensively pre-designed project let out to competitive bid. Each contractor is responsible for ensuring that the execution and standards of the building comply with the intention of the design. It is this point which is fundamentally different to the UK system.

The GSA[3] examined procurement strategies for office buildings and compared their own performance with the private sector. They found that the private sector projects were 35.5% cheaper than federal projects when comparing the costs on per m^2 of gross floor area. On analysing the reasons for this difference, the following factors were identified;

● scope – federal buildings had more in them than their private counterparts (eg interior tenant work, special facilities such as courtrooms, special features such as extra standby power, etc);
● quantitative – federal projects required more quantities of materials and components to enclose the same given floor area, ie their plan forms and geometric layouts were more complicated than the private counterparts;
● qualitative – federal buildings demanded higher performance and better quality was specified;
● unidentified causes – not attributable to any one of the above, or which may arise due to intangible factors.

It was concluded that of the difference in capital cost the approximate apportionment in cost terms to each factor were as follows;

Scope	39%
Quantitative	21%
Qualitative	31%
Unidentified	9%
	100%

Those factors within the scope and qualitative groups were matters of specification and could be amended by changing the specification or altering the quality of the building. The quantitative aspects are probably more pertinent and contribute a difference of 7.45% of the capital cost. It is probable that the private sector was, because of its need to maximise its financial return, reducing costs by analysing every aspect of the design. The study did not show how the reductions were achieved, except to say that the spaces were packaged more efficiently and there was a reduced building material content.

The ability of US developers to achieve cost reductions is due largely to the demand for value for money. It may be argued that this is no different from the UK but it is the constructor and his sub-contractors in the US who are encouraged to seek savings by improving methods when preparing the detailed design and shop (manufacturing) drawings. This is done with the aid of value engineering/analysis (VE), the provision for which is incorporated into normal forms of contract.

Value engineering/analysis was developed after World War II primarily within the US defence departments. It was introduced into the construction industry from 1962 onwards, principally by Lawrence Miles who described it as; 'a disciplined action system attuned to one specific need; accomplishing the functions that the customer needs and wants'. The basic function is to determine waste and wasted effort and to remove it.

VE works by the contractor or designer reviewing the design and questioning every element in it to determine whether it is essential in meeting the client's brief. If it is not, or an alternative can be substituted which reduces costs, simplifies construction or enhances the performance, then a VE approval is awarded and the amendment incorporated in the design. It is normal, as an incentive, to share the benefits of the savings in cost between the employer and the contractor. VE proposals can affect many aspects of a project and the full ramifications of each change have to be presented. This calls for many skills to be input into the analysis. This is possible with the American contractual system because it places more responsibility and emphasis with the contractor and his sub-contractor. As the detailed and shop drawings are produced by the sub-contractor practical solutions to design problems are achievable and there is a single point responsibility, the contractor, for producing the detailed design.

The effectiveness of this technique has been summarised[4] in terms of potential savings as follows;

– on total budget 1-3%
– on large facilities 5-10%
– on incentive contracts 0.5-1%

The performance of the US administration in using these methods produces savings of approximately $2m per annum (1973 prices), with a 45% return on the cost of operating the programme. It is interesting that;

'They further estimate an additional return of four times this amount due to in-house programs during the design phase prior to contract award'.

which could be interpreted as a 180% return on the cost

of the effort. As a further example, on 10 projects using a professional construction manager, the contribution of value engineering was from 0.8% to 7.3% of the construction cost, with a mean of 3.8%.

In summary value engineering on US experience contributes a 0.5% to 10% reduction in the capital cost of the project, with 3% - 4% as a realistic norm. It should be recognised that the clients' request this contribution to be made via the existing contract system which is already highly competitive. However, the fact is that over 70% of a projects costs are fixed very early in the design and that value engineering during this phase could increase the return four fold, to say 14% as a conservative figure.

Continental West European practice

The second area to be investigated was mainland Western Europe, which has a different contracting system to both the US and the UK. One of the major differences, although there are variations between countries, is the lack of general or main contractors.

The construction industries operate with the architect, or in Germany with the oberbauleiter, acting as designer, co-ordinator and fiscal manager of the project. They organise sub-contractors or sub-traders to do the work. The other major difference is that the contractors and sub-contractors are expected to prepare and be responsible for, the detailed design of items such as foundations, structures, mechanical and electrical services. This is a very similar requirement to that of the US sub-contractor.

It is difficult to make cost comparisons within Europe (Skoyles[5]) the main reason being the different nature of the terminology and traditional practices. However a limited study by Goodacre[6] compares one office building using Spons 1979 for the UK figures and figures from the Institut fur Bauekonomie for the West German costs. This demonstrated a similar spread of elemental costs, except that the UK services element was significantly larger. In overall terms the mean cost of construction was 13% higher in West Germany in spite of the fact that average earnings are 40% higher and the cost of living 20% higher than the UK.

Whilst it can only be inferred that the West Germans are more efficient, the absorption of higher earnings into the construction costs must be considered. The contractual system operated within West Germany and many continental countries either requires the architect to place separate sub-contracts and co-ordinate them on a direct trade basis, or to pass the responsibility for the execution to a separate agency. Both systems have the effect of allowing those sub-contractors who design and execute to develop designs which will allow the site workforce to be as efficient as possible. It is the case that the sub-contractors' designers are designing to maximise on-site production and thus absorb the higher labour costs. The architect, therefore, has a strategic role, similar to the US architect and is not involved in developing the construction detail as an important part of his function as in the case in the UK.

The French system appears to operate in a very similar fashion. On one project[7] the normal flexibility of the indigenous sub-contractors was reduced considerably by the use of a UK architect, who insisted on designing to a very considerable degree of detail.

United Kingdom practice

A third area of research has been in the UK where the Department of the Environment[8] conducted an investigation into the rationalisation of the building process. Earlier studies showed that it takes on average 1200 man hours to complete a dwelling, which it was considered could be reduced. By a combination of rationalisation of the design and management of the workforce it was possible to reduce this to 930 hours, a reduction of 22.5%.

An unpublished study, which analysed the construction implications of the design of a building without amending any of the components or changing the specification, also concluded that there was the scope in the project to reduce its capital cost by 7%. This was done by a combination of solutions to the problems which the contractor would have faced. The main one was the careful determination of the contract period, the second was the amendment of the design of the building to simplify the construction and finally others which eased the contractors working on a restricted site.

Summary and Conclusions

Whilst there is no reliable way of assessing the total impact worldwide for designing with production in mind, there are efforts being undertaken around the world to attempt to include its contribution within the competitive bid system. When it is achieved there are demonstrable savings, probably in the range 1% - 14% which, if applied across the total UK construction industry, would mean a potential increase in construction output of between £50m and £690m per annum (1981 prices).

Therefore, by considering the operational requirements, there is the possibility of significantly improving the cost effectiveness of the industry. Whilst the evidence is not unassailable it is consistent in its message.

The most noticeable feature of the American and European research reports has been the conclusion that their operatives are more productive than those in the UK. In the comparative study between America and the UK[9] it was noted that with a wage level four times higher in the US and similar material costs, building costs were not significantly different. The conclusion is that the operatives were far more productive. The problem, therefore, is how to identify the factors which control the productivity within a particular design.

DETERMINATION AND MEASUREMENT OF CONSTRUCTION WORK

The determinants of productivity operate by the interaction of two parts, the operative and his task. Summarising the factors which affect these two parts;

The Operative

● Work place conditions
If the work has to be done in cold, wet, dirty conditions then productivity will be lower than a job in warm, dry and clean conditions. These present physical and psychological factors which inhibit the operative to produce, but they are conditions which can be ameliorated by management action.

● Operative skills
Site management must select tradesmen with the correct level of skill to achieve the specification set by the design team. Whether the correct levels of skill are available is dependent upon the market forces of supply and demand, but the UK industry as a whole does operate with a fairly highly skilled workforce. However, sufficient time to achieve quality of finish must be allowed because some tasks cannot be rushed.

● Materials
To achieve smooth, progressive production the correct material must be to hand immediately otherwise the operative will stop, either to search for the material or to wait until it is supplied. Management must, therefore, ensure continuity of supply.

● Attendance
To enable materials to be supplied continuously the operative must have appropriate backup attendance. Bricklayer gangs, for instance, work in modules of two operatives to one labourer (the 2 and 1 gang) but equally trades such as plumber, carpenter and joiner must be adequately supplied. Management must, therefore, ensure that there is sufficient labour to support all trades working to their optimum.

● Mechanical assistance
Optimum operative productivity can be achieved by using the most appropriate tool or piece of plant. Power nailing, for instance, can increase nailing productivity by 400%, but, as shown in Figure 7 the pattern of major plant use is very varied through the trades. Figure 7 plots the most likely items of plant which could be used to perform the work in each trade category. Some broad assumptions as to the work content in each trade have been made to make the table as comprehensive as possible, and consequently every item of plant is not required for each trade on every job. Some plant items can be included in the rates of sub-contractors, whether nominated or direct and are indicated by · in the chart. However, it does show that there is no consistent pattern between trade category or combination of categories. If it is assumed that building work progresses in time from demolition to painting, then

Figure 7 Plant items related to SMM6 categories.

Plant columns grouped as — **TASK TOOLS** (Screed Pump, Hydraulic Exc., Hydraulic Exc., Crawler Loader; "Normally in Rates"); **HAND TOOLS** (Vibrators, Power Floats, Vibrating Tamps, Brick Cutter, Bar Bender, Bar Cutter, Saw Bench, Saw Table, Welding Set, Breakers); **GENERAL EQUIPMENT** (Concrete Pump, Tower Crane, Mobile Crane, Batching Plant, Concrete Mixer, Mortar Mixer, Compressor, Placer (Comp. Air), Dumper, Fork Lift/Placer, Goods Hoist, Pumps). Legend: · = dot, ⊙ = circled dot.

Type of Construction	Screed Pump	Hydraulic Exc.	Hydraulic Exc.	Crawler Loader	Vibrators	Power Floats	Vibrating Tamps	Brick Cutter	Bar Bender	Bar Cutter	Saw Bench	Saw Table	Welding Set	Breakers	Concrete Pump	Tower Crane	Mobile Crane	Batching Plant	Concrete Mixer	Mortar Mixer	Compressor	Placer (Comp. Air)	Dumper	Fork Lift/Placer	Goods Hoist	Pumps
C Demolition			⊙	⊙										⊙			⊙		⊙	·	⊙					·
D Excavation & Earthwork		·	·	·										·			·				·		⊙			
E Piling & D. Walling			·		⊙				⊙	⊙	⊙		⊙				⊙		⊙		⊙		·		·	
F In-situ					·	·	·								·	·	·		·		·		·	·	·	
F Reinforcement									·	·	·		·			·	·									
F Formwork											·					·	·				·					
F Precast																·	⊙									
F Composite															·	·	·	·	·	·	·		·	·	·	
G Brickwork & Blockwork								·									·			·				·	·	
H Underpinning			·														·			·	·				·	
J Rubble Walling																	·			·					·	
K Masonry								·									·							·	·	
L Asphalt Work																	·								·	
M Roofing																	·							·	·	
N Woodwork												·													·	
P Structural Steel													⊙				⊙							·	·	
Q Metal Work													·				·							·	·	
R Plumb & M & E. Inst													·											·	·	
S Electrical Inst																	·							·	·	
T Floor & Wall Finishes	⊙				·														·	·	·			·	·	
U Glazing																								·	·	
V Painting & Decorating																									·	
W Drainage			·	·										·			·				·			·	·	·

plant is used throughout the job at different stages not necessary consecutively nor simultaneously. This presents major problems of co-ordination and efficient use by site management. Where a piece of plant is not immediately available then delays to the operative and operation will occur.

To summarise, it is the role of site management to keep the operative continuously at work, properly supplied with materials, with the proper tools available (Borcherding et al[10]). Failure to do this will result in reduced productivity.

However, there is the second part of the productivity equation to consider:

The Task
● Quality
The determination of an acceptable level of quality is a matter of arbitration between the design team, the contractor and the operative. Vague specification clauses such as; in the best common practice, to the architect's satisfaction and formwork shall have no deflection whatsoever; are commonplace and are used to obtain particularised results (Tucker[11]). But

11

with the reduction in the number of fully experienced craftmen available to the industry, a type of specification which assumes a craftmanshp's interpretation is no longer appropriate. If a particular level of craftsmanship is expected then it should be specified. This would remove the risk of uncertainty in the content of the task.

● Quantity

Whether there is a large or small quantity of work for the operative to do in a day must condition the overall productivity. If an operative can work steadily and continuously without interruptions between meal breaks then he will produce the maximum that is possible. If the supply of work fails, then an operative will stop and either; move to another work area, wait for instruction, leave the site for that day or leave the site and go to another site. In all cases there is a time between finishing one piece of work and starting the next, the variability of which is dependent upon external factors and management skill. Site management can attempt to assemble work for a particular operative skill into a consecutive series of operations. However, if a series of tasks are plotted on a line of balance, (Figure 8) the inefficien-

cy can be seen in the time buffers required between each grouping of tasks in respect of the whole project.

The smaller the quantity within each section of work the more tasks there will be and a project moves from a simple series of events to a complex one.

● Complexity

A complex task for an operative is one which calls for all of his skills and attention to produce the required finished product. The joinery item for instance in Figure 9 is complex, involving many varieties of wood and wood products built into an unusual form to a very high standard of finish. The skill with which the finish will be achieved is the craftsmanship but the complexity is in the task to be achieved.

There is a view that complexity is represented by combining a number of operations into a series. Thus, complexity is seen as the incidence of different kinds of work (Bennett and Fine[12]), which are termed work packages. As suggested above these are based on the quantity of work, which is probably a simplistic view. Bennett uses three identifiers to

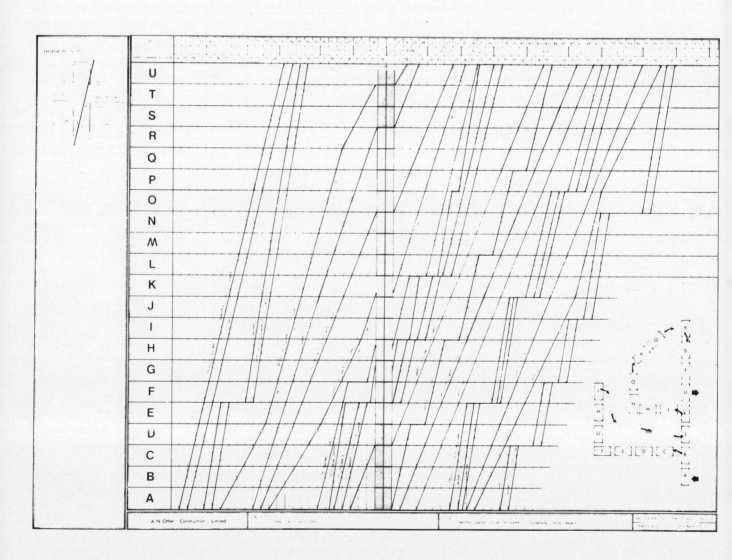

Figure 8 Typical line of balance showing delays between operations.

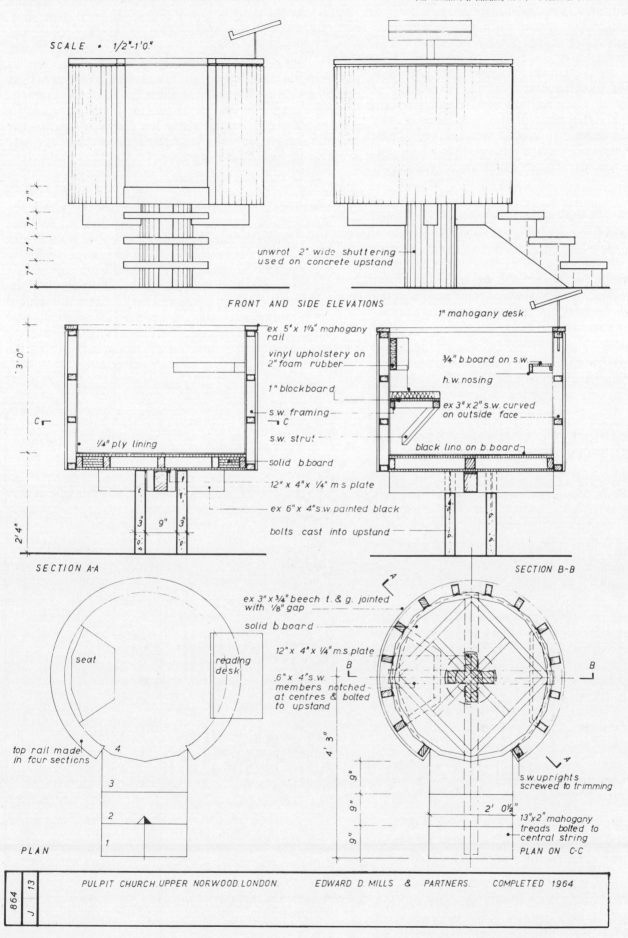

SCALE • 1/2"-1'0."

FRONT AND SIDE ELEVATIONS

unwrot 2" wide shuttering used on concrete upstand

SECTION A-A

SECTION B-B

ex 5" x 1½" mahogany rail

1" mahogany desk

vinyl upholstery on 2" foam rubber

¾" b.board on s.w.

h.w. nosing

1" blockboard

s.w. framing

ex 3" x 2" s.w. curved on outside face

s.w. strut

black lino. on b.board

¼" ply lining

solid b.board

12" x 4" x ¼" m s plate

ex 6" x 4" s.w. painted black

bolts cast into upstand

seat

reading desk

ex 3" x ¾" beech t. & g. jointed with ⅛" gap

solid b.board

12" x 4" x ¼" m.s. plate

.6" x 4" s.w. members notched at centres & bolted to upstand

top rail made in four sections

s.w. uprights screwed to trimming

13" x 2" mahogany treads bolted to central string

PLAN

PLAN ON C-C

PULPIT CHURCH. UPPER NORWOOD. LONDON.   EDWARD D. MILLS & PARTNERS.   COMPLETED 1964

864   J   13

Figure 9 Joinery items demonstrating task complexity.

select the work package;
- operationally significant locations;
- type of work;
- operationally significant functions.

Summary and Conclusions

The provision of resources, their control and the quality achieved are functions of site management. The overall performance is controlled by the task to be performed as the resources are restricted in their unfettered action by external limits which are imposed upon them.

In a perfect world, with none of the external limits such as those which affect the operative directly allowed, ie the project was managed correctly, only the parameters of the task would limit the achievable productivity. These are design parameters, the quality to be achieved, the quantity of work between physical breaks in the building and the resulting number of breaks or complexity of the sequences of tasks.

Therefore, by identifying these tasks, their scope, complexity and sequence there is a framework for identifying the productivity of the job as a whole.

TASK COMPLEXITY AND THE RELATIVE CONTRIBUTION OF SIMPLICATION

At the simplest level, work proceeds in a chain of individual tasks as shown in Figure 10

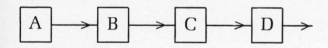

Figure 10 Simple chain of tasks

However, each task can be performed at a different rate depending on the difficulty and complexity compared with the simplest form of the task. The significance of the difficulty of a task has been analysed by access to the work study library of a national contractor. The use of work study allows analysis at a common level of 100 rating, ie a 'normal' performance level of a well motivated operative, which removes the variability of operative performance and conditions.

The range of performance, Figure 11, therefore, shows how the operative's output can vary in response to the complexity of the task, the material, the skill required and the quality expected. As these criteria can vary over a building the figures are the range of variability across the work normally undertaken by a trade.

Simplification at the task level

One view of simplification is to vary the work within a work package. Such examples are; the creation of larger bays of concrete; removing decoration on brickwork or reducing the thickness of screeds. These have the effect of increasing the rate at which the task can be performed.

To test the effect of varying the performance rate of the task, ie changing the work content due to design variation, each of the operations in a sample project was varied within the range of operation variation in Figure 11. They were modified by a random number to simulate the effect of simplifying some tasks more than others. This simulation is plotted as the lower graph in Figure 12. The results show the effect per week of simplifying the work within the task, when expressed as a percentage of the total cost of the project per week, over the period of the project.

The labour cost of the trade task varied within the range of +0 .14% to −0.01% of the contract sum per week. When the variability of all activities is taken into account the total variability is in the range 0.001% to 0.022%. The consequences of simplification of the task are either that the resources can be reduced or that the same level of resources can be used and the duration of the activity can be reduced. There is, therefore, the cost variation due to reducing the project duration to be added into the calculation but this can only be done if the activity in question is on the critical path.

Operation	Operative performance variability + or − %
Substructure	5
Concrete structure	15
Brickwork	7.5
Asphalt	2.5
Carpenter 1st Fix	2.5
Joiner	2.5
Metalwork	2.5
Structural steel	10
Builders work in connection with services	10
Plastering	5
Painting	7.5
External works	5
Piling	5
Curtain wall	7.5
Patent glazing	7.5
Rain water installation	7.5
Soil installation	2.5
Hot and cold water installation	5
Heating and ventilation installation	12.5
Lifts	10
Electrical installation	12.5
Suspended ceilings	10
Equipment	Nil

Figure 11 Operative performance variability range.

Simplification of the sequence of activity

Figure 12 also plots the value of the preliminary costs per week which range between 0.03 to 0.095% of the contract sum per week, with the peak during the structure, cladding and early finishes stages of the project. If the sequence of activities is reduced by the removal of stages in the work then the overall time of the project would be reduced. The comparison in Figure 12 shows that if a reduction in overall project time can be achieved, the cost reduction per week is greater, by a factor of 475%, than the cost of simplifying work at the task level.

Summary and Conclusions

Whilst there are undoubted benefits of simplification of the task to be performed, the benefits available through simplification of the sequence of operations are far more significant. There is, therefore, a need to be able to predict the sequence of activities in a particular design and to demonstrate the effect of removing tasks, thereby simplifying the work or simplifying the task itself.

THE PROBLEM OF MODELLING COMPLEXITY AT THE DESIGN STAGE

Any analytical and modelling technique which is suitable for use during the early stages of design must be;

- simple to use;
- capable of use by design teams;
- able to provide answers to the 'what if' question;
- able to respond from minimal information.

A comparison of the various modelling systems available, Figure 13 (Adrian and Boyer[13]), reveals that

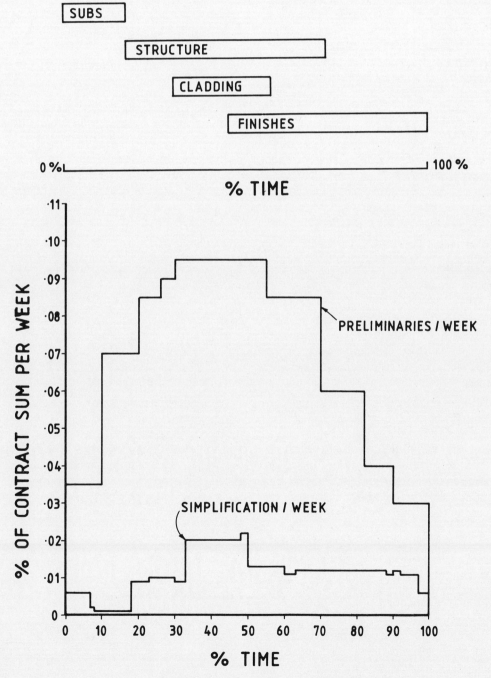

Figure 12 Relationship between work simplification and time simplification in percentage project costs per week.

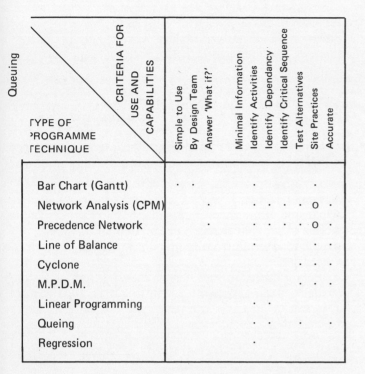

TYPE OF PROGRAMME TECHNIQUE	Simple to Use	By Design Team	Answer 'What if?'	Minimal Information	Identify Activities	Identify Dependancy	Identify Critical Sequence	Test Alternatives	Site Practices	Accurate
Bar Chart (Gantt)	·	·		·						·
Network Analysis (CPM)		·			·	·	·		o	·
Precedence Network		·			·	·	·		o	·
Line of Balance							·			
Cyclone								·	·	
M.P.D.M.								·	·	·
Linear Programming				·						
Queing					·		·		·	
Regression				·						

Figure 13 Ability of time prediction programmes to analyse work packages during the design development stage.

none possess the attributes required, particularly that of operating from a minimal level of information within design teams. All the techniques require considerable information for the calculation of the quantity of work within the activities and the subsequent duration calculation and are more suited to the stage when the design is more fully developed, usually nearer to the bid stage and too late in the process to affect initial design decisions. Therefore, new techniques of analysis and prediction are required.

Modelling complexity at the early design stage

Whilst it may be possible to model the construction operations, which will be discussed later, the value of the information must be capable of interpretation and of being used effectively. The analysis (in Figure 12), used the relationship between overall time, expressed as the value of the preliminaries and operational labour input, as the measure of the relative contribution of simplification. Money is the obvious transfer function as comparisons can be made in a common unit. Therefore, the model must produce the information required to enable an accurate calculation of the cost of preliminaries.

Conventional practices of design teams, however, involve the calculation of preliminaries by making a percentage addition to the total net value of the measured work. But the data base on which this projection is based includes the profit, overheads and method related costs of the contractor's operations. The resulting cost of the preliminaries using this method of calculation can only be sensitive to the total cost of the bill of quantity items and not to the consequences of method and time.

An extensive programme of research has been undertaken to investigate the way in which preliminaries are calculated (Gray[14]). The main conclusion is that preliminaries can be calculated from a basic set of design information and a limited set of time period information from the construction programme.

The time periods are summarised in Figure 14, which shows that through the building process there are only 18 time points required to calculate the input to the preliminaries calculation. In practice the selection of the periods to include varies with the type of construction and the building's size. A model must, therefore, enable these points to be accurately determined and the time between them to be calculated. It should also be noted that the time periods are relatively coarse measures covering fairly broad time periods and only measured to the nearest week.

This leads the analysis away from the detailed operational level to a point probably mid-way between a high level programme (master plan) and a detailed programme (weekly work schedule).

The programme must, however, respond to the complexity of the design, but only to the point of accurately reflecting the consequences of significant stages within the broad bands described above. Consequently, the scope of the work packages must be clearly identified, together with their relationship to other work packages, if the full implications of the complexity of the design are to be modelled.

Work packages

Bennett suggests that each work package can be identified from another by considering;

- the operationally significant locations;
- the type of work;
- the operationally significant functions.

This view has been tested by examining a portion of a typical construction programme for the activities from commencement of the superstructure above ground to the completion of roof level plant rooms. The section of the programme is shown in Figure 15 and the results of the analysis in Figure 16. A tick is used to show how the work package can be identified as being significantly different from the preceeding work package by using the criteria in the heading.

The programme can be split into three operationally significant locations. Each was selected as significant because each has a different combination of forms of construction, the implications of which needed to be assessed separately. All packages can be described as types of work but this is insufficient to discern between concrete in floor slab and concrete in columns. The definition of operationally significant function is based on the analysis of construction planning units[15] which classifies work as follows;

Total time		Duration (weeks)
From	To	
Project start date	Project completion date	
For each building:		
Commencement of each section	Completion of each section	
" Superstructure above		
ground	" Roof level plant rooms	
" Reinforced concrete	" Roof level plant rooms	
" Each floor level	" Each floor level	
" Cladding	" Cladding	
" First fixing	" Floor screeds	
" Finishes	" Of the building	
" External works	Project completion date	

Figure 14 Significant time periods for use in determining preliminaries

Setting out and excavation
Structure – load bearing
  Foundations
  Structural frame
  Walls
  Upper floors
  Roof structure
Structure – non load bearing
  Non structural envelope
  Non structural sub-divisions
  Services
  Builders work to services
  Finishes and fitting out
  Commissioning

Even this structure is insufficiently precise to differentiate between work of a similar nature, such as concrete work. Therefore, sub-classifications have been developed to show how one work package is significantly different from another. The additional columns of; material, trade and plant have been added to identify, within the sub-classification of work, how the packages are operationally significant. However, even these criteria dotnot separate out the components of operations which use the same materials or require the same skills. The function that these components perform is used by the contractor to assess the steps through which the building will grow. Whether the growth is vertical or horizontal can be used as criteria to differentiate between concrete columns, beams and floor slabs. Once the building has grown to its final shape and size the criteria of horizontal and vertical growth are exhausted as the remaining work is contained within the external envelope. The services and finishes therefore present a problem of analysis. Perhaps the notion of growth can be used but this time internal growth, from the structure to the surface finish. Thus, the service closest to the structure is fixed first to be covered by other services, building away from the structure, all to be covered eventually by the final surface finish.

Therefore, the original three criteria can only identify activities if used in conjunction with the sub-classifications shown in Figure 17.

The final finishing trades are, however, more closely interrelated than the structure and cladding and an analysis of project programmes has shown that there are two different sequences of completing work within a space for the two main stages of work:

carcase	finishing
ceilings	walls
walls	ceilings
floors	floors

These sequences can be deduced using the horizontal and vertical sub-classification but only with a knowledge of the behaviour of materials and working practices, particularly the need for providing access and of preserving the cleanliness and integrity of finished work.

Relationship between work packages
Further examination of the programme given in Figure 16, shows that the arrangement of the work packages falls into two patterns, sequential and simultaneous. At the very detailed level every task in a particular location, for example a room, will be sequential as the work by one trade must be completed before the next can start because of either insufficient space or volume of work. When considering programmes which incorporate many work tasks within the scope of a work package, there is a need to determine whether the relationship is truly sequential or whether there is a degree to which there can be overlap.

An analysis has been made of this particular problem by studying the critical path in a variety of construction programmes. Figure 18 shows schematically that not only does the critical path go from start to finish of whole work packages, but where the work packages are simultaneous the critical path passes through only a portion of each work package either at the beginning or at the end. To determine the basis on which the relationship between the work packages is calculated a detailed study of the workings within work packages has been made. Figure 19 shows schematically the degree of criticality of the various components of the

17

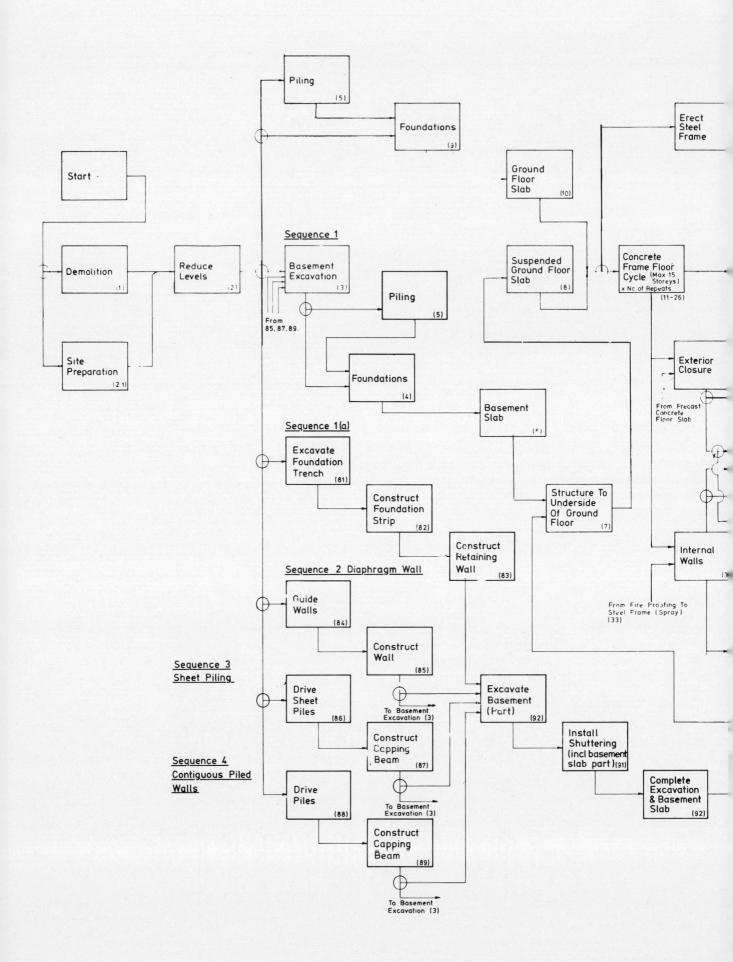

Fig 24.

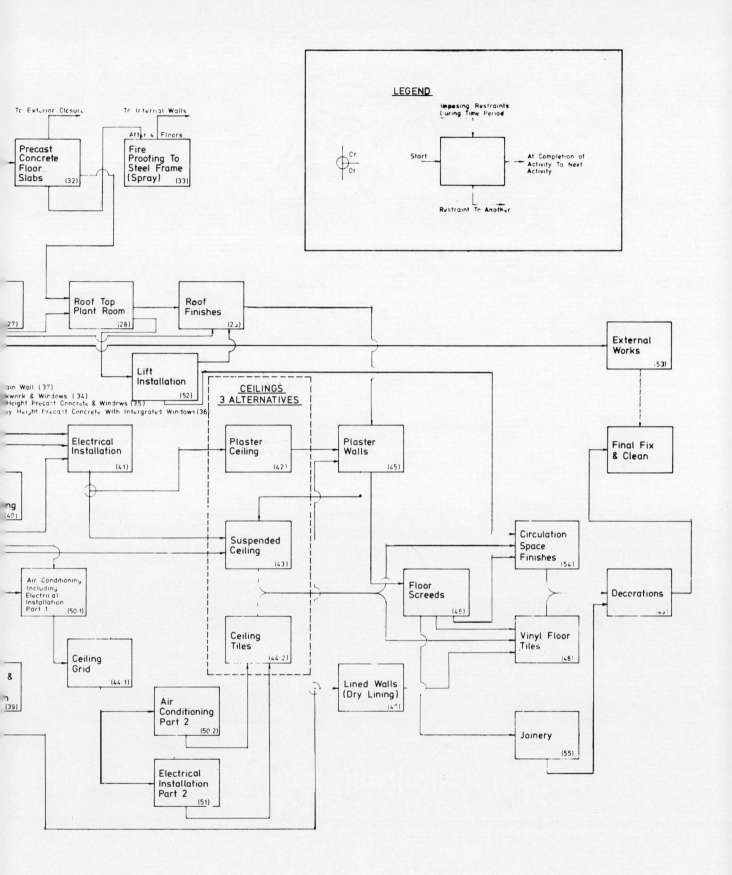

LEGEND

Imposing Restraints During Time Period

Start → At Completion of Activity To Next Activity

Restraint To Another

Cr / Cr

To Exterior Closure

To Internal Walls

After 4 Floors

Precast Concrete Floor Slabs (32)

Fire Proofing To Steel Frame (Spray) (33)

Roof Top Plant Room (28)

Roof Finishes (29)

External Works (53)

Lift Installation (52)

CEILINGS 3 ALTERNATIVES

ain Wall (37)
xwork & Windows (34)
Height Precast Concrete & Windows (35)
y Height Precast Concrete With Intergrated Windows (36)

Electrical Installation (41)

Plaster Ceiling (42)

Plaster Walls (45)

Final Fix & Clean

ng (40)

Suspended Ceiling (43)

Circulation Space Finishes (54)

Decorations (49)

Air Conditioning Including Electrical Installation Part 1 (50·1)

Ceiling Tiles (44·2)

Floor Screeds (46)

Vinyl Floor Tiles (48)

Ceiling Grid (44·1)

Lined Walls (Dry Lining) (47)

& n (39)

Air Conditioning Part 2 (50·2)

Joinery (55)

Electrical Installation Part 2 (51)

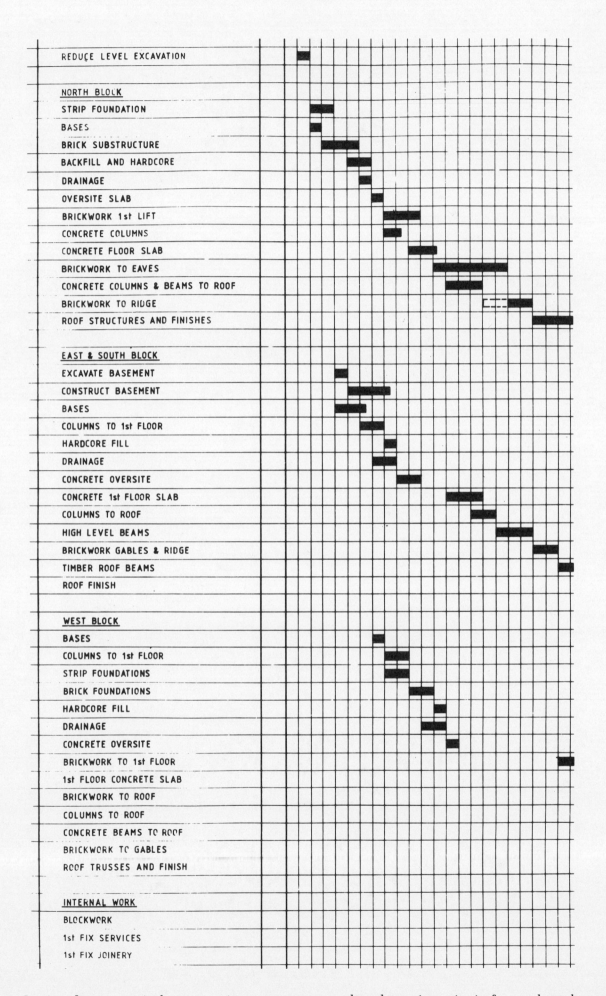

Figure 15 Section from a typical construction programme used to determine criteria for work package selection.

Operation	Location		Work			Function	
	Sequence	Size	Material	Trade	Plant	Vertical	Horizontal
North Block	*						
Oversite slab							
Brickwork to 1st floor			✓	✓			
Concrete columns			✓	✓	✓		
Concrete floor slab							✓
Brickwork to eaves			✓	✓			
Concrete columns &			✓	✓	✓		
Beam to roof							
Brickwork to ridge			✓	✓			
Roof structure			✓	✓			
East & South Block	*						
Oversite slab							
Brickwork to 1st floor			✓	✓			
Concrete floor slab			✓	✓	✓		
Columns to roof						✓	
High level beams							✓
Brickwork gables			✓	✓			
Roof beams (timber)			✓	✓			
Roof finish			✓	✓			
West Block	*						
Concrete oversite			✓	✓			
Brickwork to 1st floor			✓	✓	✓		
Concrete floor slab			✓	✓			
Brickwork to roof			✓	✓	✓		
Columns to roof							
Beams to roof							✓
Brickwork to gables			✓	✓			
Roof construction			✓	✓			

Figure 16 Analysis of activities into work packages.

Main criteria	Sub-classification
Operationally significant location	Groups of differing sequence
	Size
Type of work	Material
	Trade
	Plant
Operationally significant function	Vertical
	Horizontal

Figure 17 Operationally significant criteria for work package identification

complete work package of foundations. In this case reinforcement is the task around which the other tasks operate, based on the philosophy of selecting the key parts of the work package which control the growth of the building. Reinforcement is the most complex and slowest part of the operation as the blinding and excavation are machine orientated whilst reinforcement is labour intensive. The formwork is based on crane handled panels and is also machine oriented but cannot be finally completed until the reinforcement is complete.

The total duration of the work package can be calculated by summing the key parts. The relationship between the key parts is calculated by analysing the work space requirement of each succeeding trade. Therefore, the blinding requires a work volume released by completing 50% of the excavation. The reinforcement can commence once a work area of 25% of the blinding is complete, but once it is complete the final 25% of the enclosing formwork can be finished and the final 10% of the concrete placed. Practical allowances such as cure times for concrete must also be included in the calculation.

The relationship with another work package must, therefore, be similar. Figure 20 shows that the structure work package commences when sufficient work area is released by the completion of 50% of the foundation concrete.

It is, therefore, the speed at which the tasks of the succeeding operations consume the available work area which determine the extent of the overlap between work packages. The whole objective is to achieve a smooth uninterrupted utilisation of resources without any hold ups through work areas not being available.

Calculation of the duration of a work package

The volume of work for each task within the work package can be measured from the details of the building's design. An output rate can be determined by taking account of the degree of difficulty and quality required of the task. The duration is then determined by selecting an appropriate level of resource. The most appropriate level of resource is, consequently, fundamental to the success of the model of the whole process.

On a day to day basis it is a straightforward matter to count the areas within which each man can work for the day and arrive at the day's requirement for resources. However, when taking a wider, long term view of the resource levels within work packages the problem becomes one of determining the average level of resource, knowing full well that the site will operate on a fluctuating daily level. In the UK, with its labour employment laws, which are onerous and restricting upon the hiring and firing of employees, it is prudent to operate with a consistent labour force. Thus, operations are matched with each other to maintain an even and continuous level of resources. The contractor and his sub-contractors are attempting to minimise their costs by;

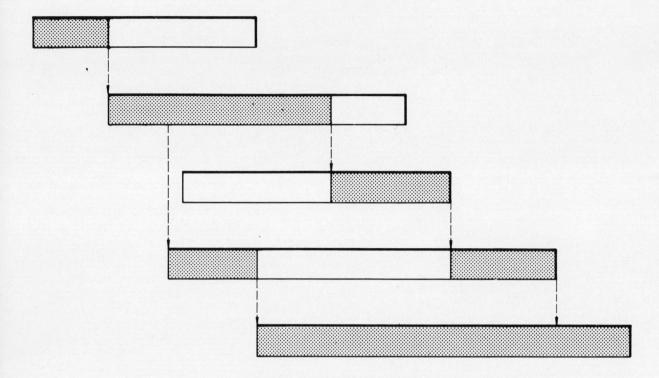

Figure 18 Movement of the critical path within activities at the strategic planning level.

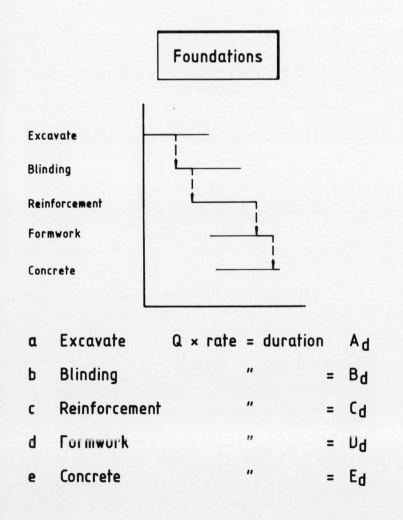

a	Excavate	Q × rate = duration	A_d
b	Blinding	"	= B_d
c	Reinforcement	"	= C_d
d	Formwork	"	= D_d
e	Concrete	"	= E_d

$$\text{Fndn} = (\cdot 5A_d) + (\cdot 25B_d) + (C_d) + (\cdot 25D_d) + (\cdot 1E_d)$$

Figure 19 Critical components of a complete work package

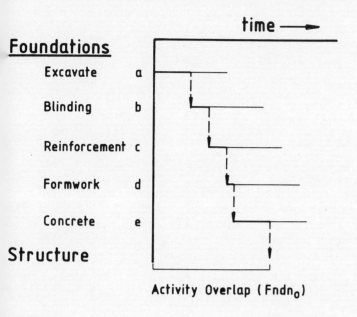

Activity Overlap (Fndn$_o$)

$$\text{Overlap} \quad \text{Fndn}_o = (\cdot 5A_d) + (\cdot 25B_d) + (\cdot 25C_d) + (\cdot 1D_d) + (\cdot 5E_d)$$

Figure 20 The critical relationship between work packages

- maximising the use of the learning curve of each gang;
- minimising of the interruption in work sequences on each cash flow of similar tasks;
- minimising any relocation of manpower once a trade has begun work;
- minimising control and supervision costs;
- placing adjacent to each other as many gangs as is feasible within the context of efficient construction[16].

As a consequence there is a rhythm built up on the site, the pace of which is set by key work packages.

The pace setting work package, shown in Figure 21, requires the tasks in work packages A, B, C and D to be complete and thus release the work area for it to start. It then controls, by the release of subsequent work areas, the start and the work rate of work packages E, F and G. Examples of this would be; the steel frame of a factory, or the floor cycle in a concrete frame or plastering and floor screeds in the finishes of most buildings. As these are the key work packages which set the pace of the whole programme, a further detailed study has been made of the way in which their durations are calculated. It is a simple matter if there is a resource of one unit which could be identified as the key resource within the work package. In the case of the steel frame and concrete frames there is the single resource of the crane. If decisions, upon which the selection of the crane is made, could be rationalised then the selection of the key resource could be made and the pattern of activity established for the programme. This applies to activities which are controlled by an external resource, usually an item of plant. Other examples are screed pumps, concrete pumps and excavators. Some activities may be controlled by the fact that only the minimum gang size of one craftsman can work in the space available or that the relative speed of all the activities in a complex work package is determined by the normal gang size of 2 and 1, or whatever is the normal trade practice.

Example of the selection of a key resource (a crane)
Some form of lifting assistance is required when unit loads exceeds 50 kg. For lifting at low levels (up to 4.5m) forklifts and telescopic boom loaders are available, but where the building exceeds this height or the ground floor is suspended or the loads exceed 200 kg (ie more than four men can lift), a crane is required. The choice lies between a mobile crane and a tower crane.

The advantage of a tower crane is its capacity, irrespective of size, to move heavy loads over a wide horizontal distance from a fixed point. It is, therefore, possible to lift substantial weights over the whole plan area of the building. This is the first criterion for tower crane selection, 100% lifting cover. The crane should also be able to pick up and deposit the heaviest loads within its normal radius of operation. If it cannot do so without becoming excessively large (ie lifting more than 2 tonnes at the maximum radius in relation to the building's extremities) then supplemental lifting equipment may be considered. Shapiro[17] summarises the main requirements as follows:

'The design of a tower crane installation includes several elements besides support of the loads. First, the crane must be positioned where there will be adequate space to lay out the crane components before erection work. When erected, cranes must be free to weathervane, needing 360° of clear space without obstruction to jib slewing. Furthermore, when in position the crane must provide hook coverage and adequate load capacity at all required points. Finally, at the completion of

23

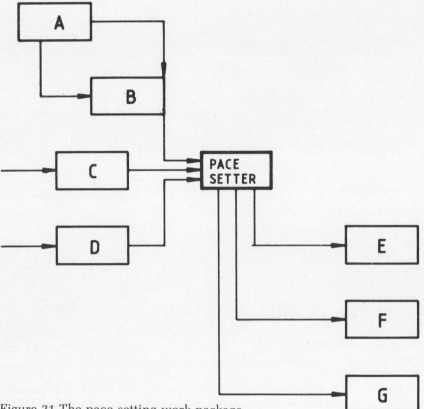

Figure 21 The pace setting work package

the work there must be access for a mobile crane or other means to disassemble the crane'.

Given the wide range of tower cranes that are available; those capable of lifting 2 tons at 15m radius with a 30m maximum radius, to those which can lift 2 ton at their maximum radius of 63.5m, there is probably one crane which could provide 100% lifting cover. This would satisfy the first criterion above but may not be sufficient to lift all the required materials in the preferred programme time. A method of checking whether there is sufficient time to lift the loads is shown in Figure 22.

This calculation sheet is used for each work level. The quantities of a given design are converted into loads by using the divisor in column 2 which is then multiplied by the time to lift each load in column 3. The total time the crane is required to work is the sum of all the load lifting times plus a contingency allowance and an allowance for delays due to weather.

Assuming an eight hour working day the number of crane days can be calculated.

From the basic data of crane days, decisions can then be taken on the following;

(a) if one crane is used, whether the floor cycle time for the structure will be the same as or greater than the minimum crane days, or whether the speed of the project is controlled by the crane;

(b) if time is of the essence, more than one crane may be required, thus releasing work areas for more than one set of resources;

(c) if several cranes are required, given the available

space and positioning, but are insufficient in terms of total crane days then supplementary equipment, such as concrete pumps or high speed concrete hoists can be introduced.

This calculation can also be made for mobile cranes but an addition of 15% to the lifting times must be made to allow for the slower work rate of these machines when compared to tower cranes. Whilst there are many factors which must be considered when selecting cranes, these basic principles can be used to measure the number of cranes required on any project. Therefore, the key resource can be established using the decision flow chart in Figure 23.

Summary and Conclusions

The design of a building can be analysed into the number and series of work packages necessary for its construction by using a simple set of decision criteria. Once the work packages have been identified they can be used to model the construction process by establishing the linking relationships between them. It is possible to locate the key or pace work packages which control the pace of the production of the preceeding and succeeding work packages, by recognising the need to maintain continuity of resource utilisation.

This structure of relationships can then be used to generate a model of the construction programme.

By the careful analysis of a design into its constituent work packages it is possible to test the construction implications of alternative design strategies by comparing the total implications of the alternative work packages

Work Item	Quantity	Load Size	Lifting Time (Stand.Mins)	Total (Standard Minutes)
	(1)	(2)	(3)	$\dfrac{(1) \times (3)}{(2)}$
Concrete				
Volume in slab		.64m³	4.5	
Volume in beams		.64m³	6.0	
Volume in columns		.29m³	5.0	
Volume in walls		.29m³	3.5	
Formwork				
Area of beam sides		4.0m²	5.0	
Area of column sides		3.0m²	5.0	
Area of wall sides		3.0m²	15.0	
Area of soffit) deck x .348		15	5.0	
)support x .1		20	5.0	
)props x .665		20	4.5	
Reinforcement				
Area of slab		14m²	4.5	
Area of beams		4m²	5.0	
Number of columns		l No.	3.0	
Area of walls		3m²	5.0	
Sundries				
Waffle moulds		8 No	5.0	
Trough moulds		8 No	5.0	
Hollow poots		60 No	5.0	
Precast units		1 No	14.0	
Scaffold perimeter length		5m	4.5	

Add Contingency 10%

Add weather 10%

$\div$ 60 $\div$ 8 = 'crane days'

Figure 22 Crane days calculation sheet

and their combinations. This, however, requires an ability to select the work packages and calculate their relative importance as the design is forming, probably even before it is committed to paper, because as soon as a design starts to be drawn it is physically and psychologically extremely difficult to amend and the commitment to cost is made.

NPS A SYSTEM FOR MODELLING THE DESIGN OF BUILDINGS IN OPERATIONAL TERMS*

Overview of the development of an alternative modelling system

Using the concept of work packages a system has been developed, the Network Processing System (NPS), to model the construction process. It differs significantly from existing computerised planning systems such as CPM in that it is designed to operate with a minimum level of detail of the particular project and yet give an accurate forecast of the likely time to construct the various parts. This cannot give absolute precision of every activity in the project but it is sufficient to enable information on the 18 time points to be given for input into the cost calculations. The system requires that the work packages are identified and that the relationships between activities are stated as relationships and not fixed periods. This releases the major constraint of the existing systems which must be tailored specifically to the exact situation to be modelled. This concept is more fully explained in the section on link transfer expressions below. To be wholly effective a system for modelling construction operations must be able to function as soon as the design has started to be formulated. This could be within stage A of the RIBA plan of work (inception) or certainly by stage B (feasibility) where the

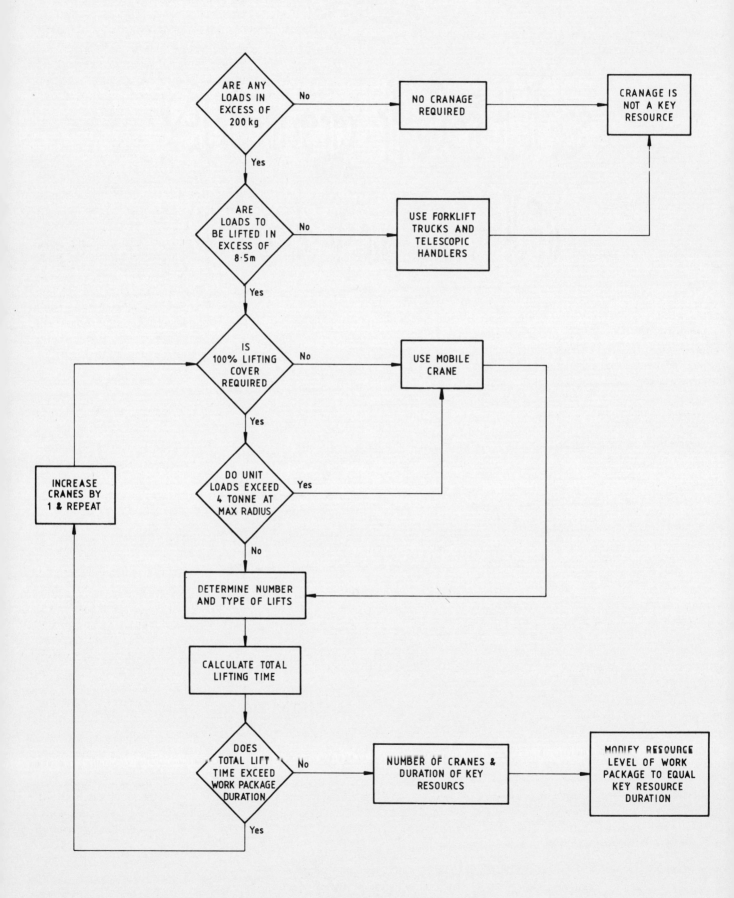

Figure 23 Decision tree for the selection of a key resource

client's requirements are already required to be costed. It is exceptional to have a contractor involved at this stage, therefore, any system must incorporate the contractor's expertise but operate within the design team environment.

As the system is designed for use by design teams the following criteria were established for its development. That the system be;

- simple, economical and quick to use;
- flexible, to meet continuously changing criteria and conditions;
- capable of giving the designer a measure of the impact upon the constructoin duration of alternative designs;
- capable of being easily updated;
- able to use only the input measurement information which is normally required for cost assessment.

The system has, therefore, been designed to give a much freedom to the user as possible and to enable a choice of approaches. Option 1 is to use the pre-generated models of typical types of buildings (called the default set), whereas option 2 is to generate an analysis of a specific problem and use the system in a manner similar to the conventional network programs. Whichever option is chosen there is a capability of rapidly assessing alternative designs. The first stage, however, is to consider the model.

Constructing the logic model

One model to cover all types of construction is not a practical proposition due to the infinite variety of designs. However, when building types – offices, factories, shopping centres or garages – are examined the function of the building limits the range of forms and construction solutions applicable to each. This has been used as the basis to develop models for particular types of building form. Within each building form there is a range of design options to be catered for, which would affect the sequence of operations, such as;

- a basement;
- a variation in floor area;
- mezzanine floors;
- double height storeys;
- roof top plant rooms;
- lift shafts.

The system must allow a choice to be made in order to build a model which accurately reflects the design. Furthermore, for each design requirement in a building there are alternative forms of construction, one of which will be satisfactory, given the particular circumstances of the project. For example, a basement can be constructed in a variety of ways, depending largely upon the soil conditions and the loads imposed by the adjacent structures;

- open cut excavation, in situ concrete retaining walls;
- propped excavation;
- cantilevered sheet piling;
- continguous bored piling;
- diaphragm wall.

A model which incorporates these basic construction operations is similar to the standard networks in existing CPM systems but the existing systems are specific, whereas NPS is a generalised computer system which allows interactive manipulation. Models for offices and factories have been established as shown in Figure 24 (factory model) which are based on the philosophy for the selection of work packages, the calculation of their durations and the calculation of the relationship between them developed earlier.

Operation of NPS

By being interactive the user can communicate easily with the program which prompts the user with a series of questions. The first step is to select which type of building is to be simulated. This has the effect of bringing forward the default model for that building type. The second step is to establish the size, shape, number of floors and the general parameters of the project. These two steps are merged and the system then takes the user through a series of questions to establish which of the construction features are required to correctly model the proposed building. Commonly, this has the effect of selecting approximately one third of the available work packages.

The system then takes over and calculates the durations and overlaps of all the work packages and relationships in the condensed model, using the basic building parameters of gross area and production calculations (m^2 of gross area output per time period) to produce the construction programme for the particular building project. This is the simplest use of the program, the advanced features of which are described later.

Features of NPS
Size of model
The model as defined to NPS is a NETWORK of previously defined logical paths, made up of ACTIVITIES (previously defined as a work package) and LINKS. Each activity is a task to be completed and the links define the inter-relationship that each activity has with the others. The entire network is based on the precedence diagram format, ie the activities within the network are the important external consideration and the links are inherent within the structure. Up to 600 ACTIVITIES can be considered in a network.

Specification of an ACTIVITY
NAME
All items within the network are known by names, ACTIVITIES need not be in any specifically numbered order with regard to their position. All activities may have more than one name, for example a reference number for quick access.

FUNCTIONS
Activity FUNCTIONS is a set of characteristics that are defined for each activity. Each function describes how

the activity will behave when the network is processed.

This behaviour is in the form of a set of expressions, one of which is chosen to represent the activity for its respective characteristics.

The TIME FUNCTION allows up to eight alternative duration calculations to be held for each activity. Thus for PILING each TIME function could accommodate the types of piling available, for example;

- frankipile;
- shell pile;
- auger pile;
- contiguous pile;
- precast concrete pile;
- steel caisson pile;
- steel H section pile;
- timber pile

With eight possible expressions per activity, different expressions can reflect different processes or resource constraints.

The COST FUNCTION is a set of eight expressions which determine the cost of the activity. Again, the expressions can be used to reflect the cost of different practices with regard to the completion of the activity, for example acceleration.

LINK
A LINK is the part of an activity's definition which determines its inter-relationship with the other activity within the network. There are two categories of links. BACKWARD links to indicate which activities preceed the activity and FORWARD links to indicate those activities which follow. To enable the network logic to be developed for particular versions of the model three types of link are available;

Mandatory link
A MANDATORY LINK, Figure 25 is one which is always active. Once the current activity is satisfied, the path automatically continues to the next activity.

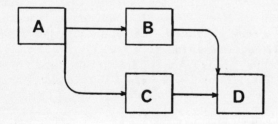

Figure 25 Mandatory link between activities

Optional/end link
An OPTIONAL/END LINK, Figure 26, is one which may or may not be active. If the link is activated the path continues otherwise the path is stopped completely. There is no need to connect ends terminated to preserve network logic and thus processing integrity.

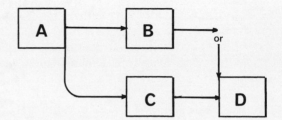

Figure 26 Optional end link between activity B and D

Optional/choice link
An OPTIONAL/CHOICE LINK, Figure 27, is one where a set of alternative activities exist of which one must be activated.

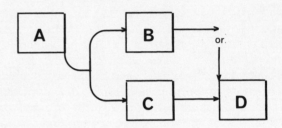

Figure 27 Optional choice between activity A and activities B and C

Wherever there is a choice presented by an option a decision must be made and those are called DECISION POINTS.

LINK TRANSFER EXPRESSION
This is probably the most powerful feature of NPS and sets it apart from other network systems. Each of the three forms of link can be assigned a value in the form of an expression which describes the relationship between predecessor and successor activities. For example;

TIME x0.3 allows the following activity to commence once 30% of the activity is complete, or

AREA x0.25 allows the following activity to commence once 25% of the floor area is available for the next task to work in, or,

a particular time delay can be set, for example, a curing time of three days of three time periods.

In this way the total time period for construction can adjust, in accordance with the needs of working practices and according to the scale of the project

This very briefly describes the principal features of the NPS system but it cannot show the full flexibility and operation, except that it is possible to model a design quickly by selecting from the many options available within the total model. By successive selection and runs, which can be developed from previous runs, many alternative design and execution formulae can be tested very rapidly as the processing time is virtually instantaneous.

BARCHART

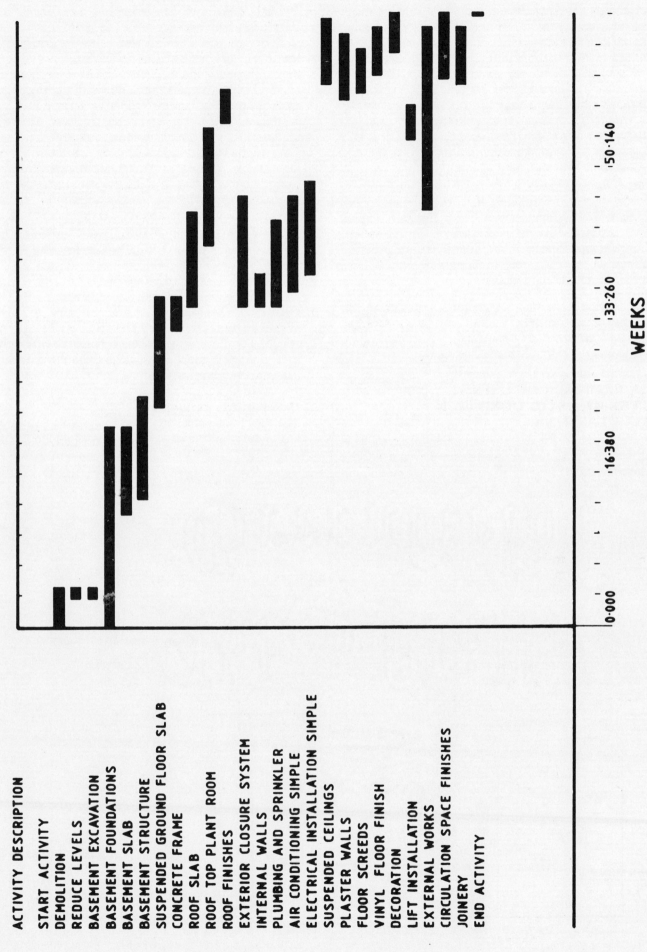

ACTIVITY DESCRIPTION

START ACTIVITY
DEMOLITION
REDUCE LEVELS
BASEMENT EXCAVATION
BASEMENT FOUNDATIONS
BASEMENT SLAB
BASEMENT STRUCTURE
SUSPENDED GROUND FLOOR SLAB
CONCRETE FRAME
ROOF SLAB
ROOF TOP PLANT ROOM
ROOF FINISHES
EXTERIOR CLOSURE SYSTEM
INTERNAL WALLS
PLUMBING AND SPRINKLER
AIR CONDITIONING SIMPLE
ELECTRICAL INSTALLATION SIMPLE
SUSPENDED CEILINGS
PLASTER WALLS
FLOOR SCREEDS
VINYL FLOOR FINISH
DECORATION
LIFT INSTALLATION
EXTERNAL WORKS
CIRCULATION SPACE FINISHES
JOINERY
END ACTIVITY

0·000   16·380   33·260   50·140

WEEKS

Figure 28 Typical bar chart output from BPS run

29

Output

As shown in Figure 28, a conventional bar chart output is produced enabling the earliest start, earliest finish dates and activity durations to be read off for each activity. No matter how complex the network, the duration information to input into the preliminaries calcuations is available in a very simple form.

Summary and Conclusions

The simulation system developed has been designed specifically for the construction industry which is task predominant, not time dominant as in other simulation systems. It is also designed as a simple tool which is transparent to the user but its power is in the ability to model complex interactions of work tasks and work packages. The ultimate power lies in the logic model, which in time must be composed of the formalised understanding of the basic activities in the construction industry, as there is rarely anything new in buildings, just their combination together.

---

*This section is written in conjunction with Dr. R. Flanagan and describes a computer program being developed within the Department of Construction Management at the University of Reading.

THE PROBLEM OF USING DETERMINISTIC DURATION CALCULATIONS

The problem of using the majority modelling systems is that the duration calculations assume a single activity output based on the most likely rate of output given a reasonable resource level. However, it should be recognised that there are many ways in which individual tasks can be performed and ideally these should be taken into account. The performance can be slow or fast and when viewed from the central point, the mean, are termed the pessimistic and optimistic output expectations respectively. However, if care has been taken to calculate each duration on a reasonable basis it could be expected to be performed at the programmed rate. Because construction is largely predesigned using well known construction techniques, this use of the deterministic duration has been regarded as acceptable. Variability does still exist but probably due to the complexity of calculation and a reluctance to show a pessimistic programme, the deterministic output calculation is used in preference to programme systems such as PERT, which attempts to incorporate variability. Within the construction industry, which deals with very complex structures and organisations, the use of deterministic outputs, however, has failed to predict events with costly consequences[18]. Techniques which attempt to predict events, particularly at such an early development stage should, therefore, consider the problem of variability of performance of the work packages and the influence of external circumstances.

Deterministic durations

The PERT formula:

$$\frac{o + 4m + p}{6} = d$$

o = optimistic duration
p = pessimistic duration
m = most likely duration

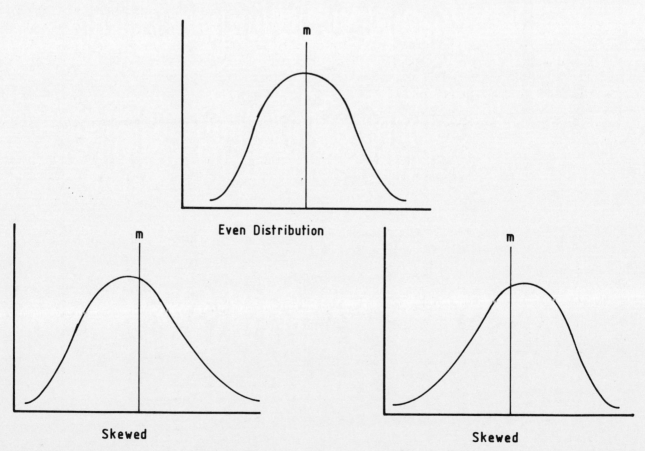

Figure 29 Skew of distribution using PERT formula

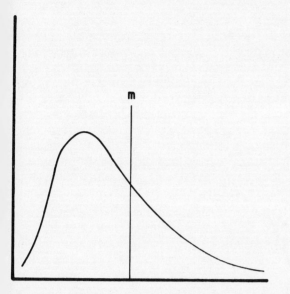

Figure 30 Skew of distribution using Lichtenberg formula

attempts to incorporate the possibility that the most likely view of the duration is not the only view.

The PERT technique was developed for research projects where the times were estimates based on guess work or best estimate with the consequences that the duration can be skewed either left or right as shown in Figure 29. Lichtenberg[19] has shown that in practice the duration is skewed to the right or pessimistic duration as shown in Figure 30. He proposes a different version of the PERT formula which will always skew the distribution to the right;

$$0.2\,o + 0.6m + 0.2p = d$$

Providing that the range of optimistic and pessimistic durations is available then this technique has considerable potential of simplicity in use. Recent experience on petrochemical projects has shown that the range of optimistic to pessimistic is between 1 to 3 and 1 to 5. However, there are other factors which must be considered in addition to the variability of the work package duration.

External factors affecting groups of work packages

Particular groups of activities may be affected by circumstances which produce uncertainty only for that group, for instance, soil conditions will affect only the substructure work. Lichtenberg suggests that allowances should be made for the following as typical example reasons;

- scope of the work;
- inconvenience;
- particular production procedures;
- climate;
- labour and skill availability;
- quality of information;
- management capability.

In studies, allowances for these particular circumstances add between four and six months to projects lasting up to four years. As these circumstances affect particular groups, then compensation for their effects can be made by adding in, at the end of the group, a balancing activity, as shown in Figure 31.

Merge event bias

The main problem noted by users of network systems, even those incorporating a PERT evaluation of work package durations, results when the process ignores paths other than the mean critical path when evaluating the probability of completion. The effect is to give an over optimistic view of the probability of completion. Klingel[20] observed that forecasts tended to underestimate actual completions by a third of the total period. This has been called merge event bias and occurs where several paths converge on a single network event. One of the paths will be the longest net path but in practice may not be the earliest start time of the single event because variability in the completion of the other paths may exceed the original critical path. An example of the point of merge even bias is also shown in Figure 31. Van Slyke[21] suggests that the use of Monte Carlo simulation would enable a solution to be obtained which accounts for this problem but in his experiments over 10,000 iterations of the complete network were used to arrive at a suitable schedule, which for practical reasons is unsuitable for general use.

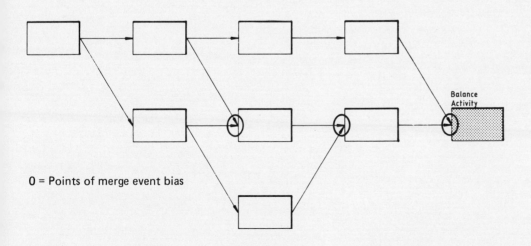

0 = Points of merge event bias

Figure 31 Location of merge event bias and the use of a balancing activity

Crandall has refined this technique, but is still using the Monte Carlo simulation to generate the variability within the performance of the activities. This gives a more random answer than Lichtenberg's whose view is that the bias should be pessimistic. The following technique is a development of Lichtenberg's work. By incorporating a Monte Carlo type simulation into the duration calculation of each activity, but within a beta distribution framework established by Lichtenberg's amended PERT formula, it would be possible to establish the most likely duration and the variability necessary to evaluate the scale of adjustment required to account for the merge event bias. Most modern computer systems include these routines as part of their standard software.

Therefore, using the beta distribution (Johnson and Kotz[22]).

$$P(x) = \frac{1}{B(p,q)} \frac{(x-a)^{p-1}(b-x)^{q-1}}{(b-a)^{p+q-1}}$$

$$(a \leq x \leq b); \ p, q > o$$

a = the optimistic duration
b = the pessimistic duration
p,q = parameters of the distribution; p, q o
B(p, q) = beta function

With a suitable number of iterations the most likely duration of the activity can be calculated. The problem, however,is the determination of the optimistic and pessimistic durations. The deterministic duration can be calculated by;

$$\frac{m}{r \times n} = d$$

m = the measured quantity
r = the resource quantity
n = the production output

Unfortunately, there are no established limits of the optimistic and pessimistic range but it is unlikely that the work package will be completed in half the time assessed but perhaps 10%-15% quicker than the calculated duration. Even this judgement, compared with the analysis of programmed to actual completion (Roderick[23]), is optimistic. Only one activity in Roderick's study was completed within the predicted time. 76% of the activities were completed within 500% of the optimistic time and 86% of the activities were within 600% of the optimistic time. Therefore, the deterministic time could be said to be the optimistic time. The input to the beta distribution would, therefore, be;

a = the deterministic duration
b = either 3(a), 5(a) or 6(a), depending on the view of management control
p = 3
q = 2

Using these values of p and q will give a beta distribution skewed in accordance with Lichtenberg's findings. Having identified the parameters of the distribution, the mean value of the duration can be generated using a random number generator. If this is repeated for N simulations of the work package and the average and variance is calculated this is a simpler approach than simulating a network many hundreds of times, even for a computer. A normal network forward pass calculation is performed until the first converging point (event or activity having two or more predecessors). The early finish time of each path is a summation of the activity mean values. The attached variances are established in a similar way as the sum of the variances of previous activities along the path and the total variance is added to the earliest start time of the critical path to allow for inevitable variation in performance of the work packages.

In cases of long preceeding sequences of work packages, the variances should be slightly reduced by the weighting procedure of $v_k = 0.7 \ (v \ near) + 0.3 \ (v \ distant)$ where $v \ near$ is the shortest proceeding series of activities and $v \ distant$ is the longest series of preceeding activities. The network calculation is continued as described above, until the next merge event appears.

Summary and Conclusions

The result of this procedure will, for any network, be an approximately correct value of the project duration.

The Lichtenberg technique has enabled the development of results to within 0.8% of techniques using Monte Carlo simulation. Thus, by using a very simple approach to incorporate probability within the calculation of work packages from a deterministic calculation, a much simpler technique, more closely allied to practice can be developed. This, at the moment, is a theorectical approach but it does give the possibility of the development of relatively simple but accurate enhancements to the existing time prediction techniques.

References

1. HIGGINS, G. and JESSOP, G.Interdependence and uncertainty. a study of the building industry. 1966. Tavistock Institute.

2. BARRIE, D. S. and PAULSON, B. C. Jr. Professional construction management. McGraw Hill. 1978

3. GENERAL SERVICES ADMINISTRATION. A comparison of Federally and privately constructed office buildings. 1070.

4. BARRIE, D. S. and MULCH, L. The professional CM team discovers value engineering. *ASCE Journal of the Construction Division* 1977 September, pp423-435.

5. SKOYLES, E. R. The quantity surveyor and his influence on Europe. *Quantity Surveyor* 1979 January, pp345-352.

6. GOODACRE, P. et al. Britian and Germany – a comparison. *Chartered Quantity Surveyor* 1979 August/September, pp16-19.

7. Playing the field. *Building* 1980 November 14, pp34.39.

8. FORBES, W. S. and STJERNSTEDT, R. The Finchampstead Project. Current Paper CP 23/72. Building Research Station. 1972.

9. UNIVERSITY OF READING. Department of Construction Management. UK and US construction industries: a comparison of design and contract procedures. Royal Institution of Chartered Surveyors. 1976.

10. BORCHERDING, J. D., SEBASTIAN, S. J. and SAMELSON, J. J. Improving motivation and productivity on large projects. *ASCE Journal of the Construction Division*. 1980 March, pp73-89.

11. TUCKER, R. Quality in construction. Paper presented to the luncheon discussion club, University of Reading. 1981.

12. BENNETT, J. and FINE, B. Measurement of complexity in construction projects. SRC Research Project GR/A/1342.4, Final Report. 1980 April., Department of Construction Management. University of Reading.

13. ADRIAN, J. J. and BOYER, L. T. Modelling method productivity. *ASCE Journal of the Construction Division* 1976 March, pp157-167.

14. GRAY, C. Analysis of the preliminary element of building production costs. MPhil Thesis submitted to the University of Reading. 1981.

15. CRISP, P. J. Proposals for the presentation of location information in bills of quantities. Discussion document submitted to SMM Development Unit. (unpublished).

16. BIRRELL, G. S. Construction planning – beyond the critical path. *ASCE Journal of the Construction Division* 1980 September, pp389-407.

17. SHAPIRO, H. I. Cranes and derricks. 1980. McGraw Hill.

18. CRANDALL, K. C. Probabilistic time scheduling. *ASCE Journal of the Construction Division* 1977 March.

19. LICHTENBERG, S. and MOLK, L. B. Three types of bias in scheduling – and solutions applicable in practice. Proceedings of the 6th Internet Conference, 1979, pp247-262.

20. KLINGEL, A. R. Bias in Pert project completion time. calculations for real networks. *Management Science* 1966 13 December, ppB4-201.

21. VAN SLYKE, R. M. Monte Carlo methods and the PERT problem. *Operations Research* 1963 11 September, pp839-860.

22. JOHNSON, N. L. and KOTZ, K. Continuous univariate distributions. Vol 2. 1970 Houghton Miffin & Co. pp37-51.

23. RODERICK, I. F. Examination of the use of critical path methods in building. *Building Technology and Management* 1977 March, pp16-19.

ACKNOWLEDGEMENTS
I would like to thank the many people who have been involved in this project; the Chartered Institute of Building's Silver Jubilee Scholarship committee, for enabling this work to be done; Peter Harlow at the CIOB, for his patience in putting up with the often over optimistic progress reports; Professor Bill Jepson, for his contribution to the ordering of my thoughts and editorial abilities on the final draft; my colleagues at the University of Reading particularly Professor John Bennett and Dr Roger Flanagan for their perception and ability to ask very demanding questions; to all the people within the construction industry particularly Clive Tassie of Wates Construction for supplying the detailed information to determine the benefits of the contractor's contribution, and finally my thanks must go to the unending patience of the secretaries within the Department of Construction Management bent over a hot word processor.